Everyday Mathematics®

The University of Chicago School Mathematics Project

Mathematics at Home
Book 2

 **Wright Group**

The McGraw·Hill Companies

Wright Group

Send all inquiries to:
Wright Group/McGraw-Hill
P.O. Box 812960
Chicago, IL 60681

ISBN 0-07-604520-X

16 17 18 QLM 12 11

The McGraw·Hill Companies

Authors

Jean Bell
Dorothy Freedman
Nancy Hanvey
Deborah Arron Leslie*
Ellen Ryan
Barbara Smart*

* Third Edition only

Consultant

Max Bell
University of Chicago School Mathematics Project

Third Edition Early Childhood Team Leaders

David W. Beer, Deborah Arron Leslie

Editorial Assistant

Patrick Carroll

Contributors

Margaret Krulee, Ann E. Audrain, David W. Beer

Photo Credits

Covers Jules Frazier/Photodisc/Getty Images.

Contents

Introduction

This is the second of four *Mathematics at Home* books for *Kindergarten Everyday Mathematics*. Do these new activities with your child, but continue to use favorite activities from Book 1.

Always remember to let mutual enjoyment and interest set the tone, pace, and level of involvement for any of these activities. The idea is to discover the fun of doing mathematics. You can vary the numbers in the activities to suit your child's needs.

Browse through the list of children's books included at the back of each *Mathematics at Home* book. Perhaps you will find some of them on your next trip to the library or book store. Only a small number of the many books that lend themselves to mathematical discussion are listed here.

Mathematics on the Road

Road Sign Geometry

Look at the shapes of road and safety signs. Discuss the reasons why stop signs, caution signs, and information signs use different shapes, different colors and words. Use the terms *circle, octagon, diamond* (or *rhombus*), and *rectangle* to describe the shapes. Besides learning the names of the shapes, you will become more observant travelers.

3

Plan Your Trip

Discuss the route you will be traveling. About how far will you travel? How long will it take to get to your destination? You might want to look at a map or sketch your own.

Dashboards and Driving

The dashboard of a car or bus displays a fascinating collection of information. Before your child is seated, point out the gauges and encourage questions.

Look at the gauges on the dashboard together and ask questions such as:
◆ What shows the car's speed?
◆ What shows how much gas we have?
◆ Can you tell how much gas is left?

Before starting a trip, set the trip meter at zero. During the journey, have your child guess how many miles you have traveled and then read the number on the trip meter aloud. At the end of the journey, check how many miles registered on the trip meter.

A harder question is: How many miles are on the odometer? (the total number of miles the car has been driven).

How Long Is a Mile?

Estimating the length of a mile isn't easy. When driving, try to guess when you've traveled a mile. You can use a trip meter or odometer to check your guess. Major highways sometimes have roadside mile markers. Try to spot them!

Driving Around

As you travel, use position words to describe landmarks. For example you might say, "We just went *under* a bridge; We drove *over* the highway; We went *around* the lake; We are *next to* the park."

Travel Time

Use counting to measure time as you travel.
Ask your child questions such as:

◆ How long will it be until we get to the next stop?
◆ How long will the red light last?

Number Game

Look for numbers on signs, addresses, and license plates to find numbers in order. Look for 1s, then 2s, 3s, and so on.

Ordinal Numbers

Family Note

Use order words (*first, second, third,* and so on) whenever the opportunity arises. Refer to dates by order, for example: October *second* or January *sixth.*

First, Second, Third...

Line up 5 or more favorite toys. Which is *first? Second? Fifth?* Count them in order: *first, second, third, fourth,* and so on. Put the toys in a different order. Now which one is *first, second, third* . . .? Line up additional things and count them in order.

When waiting in line, perhaps in a store or at a drinking fountain, check your place in line. Are you *second, third, tenth, twelfth*? Estimate how long it will take to get to the front of the line. Think of a way to check your estimate (using a watch, counting seconds, and so on).

Counting

Count Up, Count Down

Practicing counting forward and backward will help your child become a nimble counter. Try starting at different numbers. Start at 14 and count forward. Then count backward. Count down to zero like a rocket liftoff, or count along with timers and microwaves. Try starting from 10, then the teens, and then even higher numbers.

Skip Count

Practice counting by 5s and 10s. This is called skip counting and is a useful skill.

Page Count

Occasionally count the number of pages in a book before or after reading it. Compare the count with the last page number. (However, don't let the counting interfere with the pleasure of the story!)

Mathematics Note

Good rote counting skills help children become aware of the patterns and the structure of our number system.

Ordinal Numbers in Children's Books

Read a counting book together. Invite your child to read the numerals on the page, count the objects, and predict what number will be on the next page. There are many wonderful counting books, and your child can surely find one at the local library that appeals to him or her. See page 22 for book suggestions.

Patterns

Family Note

Look for patterns all around you. The more you become aware of their presence, the more you'll see.

Musical Patterns

Listen to a pattern and then repeat it.
Clap! Clap! Tap!
Clap! Clap! Tap!
Take turns making sound patterns using your feet or hands. Try using other sounds such as Snap! Snap! Stomp!

Play some music and listen to the beat. Clap to the beat.
◆ Do you think that's a pattern?
◆ Can you combine clapping and tapping your foot to the beat at the same time?
◆ How else can you copy the pattern?

Pattern Hunt

Look around and find patterns in your home, yard, or neighborhood.

Patterns in Children's Books

Share books with patterned language, such as the many books written by Bill Martin or Eric Carle. As you read, pause and encourage your child to say the repeated parts. Ask your child how he or she knew what to say. You might explain that these books follow a *pattern* and this is why it's possible to know what to say next. See page 22 for some books with visual patterns and with patterned language.

Geometry

I Spy Shapes

Children love to play "I Spy," and they can play it anywhere. Begin with easy clues and work toward more difficult ones. For example, "I spy something that is a rectangle and has a round knob on it." Let your child "spy" shapes for you to guess.

Eating Shapes

Eat a geometric treat. Make a sandwich and decide what shapes it could be cut into. Can you make rectangles, triangles, or squares? You might cut it in half and try to nibble one half into a circle and the other half into a different shape.

Peanut Butter Balls

Yield: about ten 1-inch balls

◆ Mix about $\frac{3}{4}$ cup of crunchy peanut butter with $1\frac{1}{2}$ tablespoons of honey or sugar. (They are better if they are not too sweet.)

◆ Add about $1\frac{1}{8}$ cups (depending on the peanut butter consistency) of powdered skim milk to make a dough that is stiff enough to roll into balls.

◆ Make some big, some small, and some equal-size balls.

◆ Try some other shapes.

◆ Coat the shapes with sesame seeds if desired.

◆ Chill them, then eat!

Note: If your child is allergic to peanuts, try this activity with modeling clay for a nonedible version. Make a variety of geometric shapes.

Comparisons

More or Less

The terms *more, most, less,* and *least* are useful to understand. Compare two glasses filled with milk or juice. Which glass holds *more?* Which glass holds *less?* Then fill a third glass. Which one holds the *most?* Which one holds the *least?*

When you talk about distances, try to use terms such as *farther, nearer,* and *closer* as you make comparisons. For example, "Which is *nearer,* your school or the grocery store? Which is *farther* away, the brown sofa or the blue lamp?"

ROUGH AND SMOOTH

THICK AND THIN

WIDE AND NARROW

TIGHT AND LOOSE

FAST AND SLOW

NEAR AND FAR

LONG AND SHORT

LOW AND HIGH

Smaller than a Snowman?

If you live in an area where it snows, help your child build a snowman and compare its size to his or her size, to a friend's size, or to the sizes of members of your family. Ask: Is it *taller* or *shorter* than you? Than your friend? Is it *thinner* or *fatter* than your dog?

Ordering Items by Size

Arrange various objects (such as books, boxes, or cans) by various size attributes, such as length, weight, and volume. Talk about your arrangement using comparison words such as *taller, shorter, longer, narrower, wider, heaviest, lightest, more, less,* and *about the same.*

Estimation

Take a Good Guess

Estimate and then count how many books (or CDs or movies) are on a bookcase (or on one shelf of a bookcase). If you shared the books with everyone in your family, how many do you think each person would get? Try it and see.

Estimating with Oranges

Do all oranges have the same number of seeds? About how many seeds do you think an orange has? Estimate and then keep track as the family eats some oranges over a few days.

Do all oranges have the same number of sections?

Compare the number of seeds or sections and use mathematical terms such as *more, less, fewer, about the same,* or *equal.*

Getting a Fair Share

Sharing is easier if children are active participants in decisions that are made. For example, "We have one orange, but there are three of us. How can we divide the orange so we all have almost equal amounts?"

Problem Solving

Family Note

How can you help your child become a good problem solver? Take advantage of everyday situations to enlist your child's help in figuring out solutions to real-life problems.

Let's Figure It Out

If you have to go to the library, the grocery store, the park, and the post office, discuss where you would go first. How should you decide?

You have 2 birthday hats, but you need more. Think about how many more you need so all 5 children will have a hat.

Using Numbers to Solve a Dispute

Numbers can help settle disputes. For example, sometimes children argue about who should go first when playing a game. One solution is to have them guess a mystery number between two other numbers. For example, "I'm thinking of a number between 1 and 12." The person with the closest guess may go first. Talk about this solution: Is it fair? Why? If not, why not? Do they have other suggestions for solving problems fairly?

Mathematics Note

It is important for children to discuss their ideas and to give reasons for their conclusions. As you listen, encourage your child's thinking and accept all ideas. The discussion is what is most important.

Collecting Data

Food Findings

Together with your child, take a look at your food shelves. How many kinds of cereal do you see? How many kinds of soup do you have? Are there different kinds of canned foods? What else do you find? What does this tell you about the things your family likes to eat?

You might want to organize the food into groups on your shelves.

Family Food Survey

Children like to be involved in making food choices for family meals. What are some of your family's favorite foods? Take a survey to find out.

hamburger ‖

macaroni and cheese |

spaghetti and meatballs ‖|

Some Books for Children

Counting

Anno, Mitsumasa, *Anno's Counting Book* (HarperCollins, 1977)

Bajaj, Varsha, *How Many Kisses Do You Want Tonight?* (Little, Brown and Co., 2004)

Carle, Eric, *Rooster's Off to See the World* (Aladdin, 1999)

Carter, David, *How Many Bugs in a Box?* (Little Simon, 1988)

Feelings, Muriel, *Moja Means One: A Swahili Counting Book* (Smith Peter, 1996)

Hoban, Tana, *Count and See* (Simon & Schuster, 1972)

Scarry, Richard, *Richard Scarry's Best Counting Book Ever* (Random House, 1978)

Tafuri, Nancy, *Who's Counting?* (HarperTrophy, 1993)

Walsh, Ellen Stoll, *Mouse Count* (Voyager, 1995)

Wood, Audrey and Don, *Piggies* (Voyager, 1995)

Patterned Language

Adams, Pam and Simms Taback, *There Was an Old Lady Who Swallowed a Fly* (Child's Play International, 2000)

Martin, Bill, Jr., *Brown Bear, Brown Bear, What Do You See?* (Henry Holt, 1996)

Williams, Sue, *I Went Walking* (Gulliver Books, 1990)

Multiple authors, *The Gingerbread Man*

Multiple authors, *The Little Red Hen*

Problem Solving

DePaolo, Tomie, *Pancakes for Breakfast* (Voyager, 1990)

Hutchins, Pat, *Changes, Changes* (Aladdin, 1987)

Lobel, Arnold, *Ming Lo Moves the Mountain* (HarperTrophy, 1993)

McDermott, Gerald, *Arrow to the Sun: A Pueblo Indian Tale* (Viking Press, 2004)

McGovern, Ann, *Stone Soup* (Scholastic, 1989)

San Souci, Robert, *Six Foolish Fishermen* (Hyperion, 2000)

Spatial Relationships

Berenstain, Stan and Jan, *Bears in the Night* (Random House, 1971)

Carle, Eric, *The Secret Birthday Message* (HarperCollins, 1974)

Hennessey, B.G., *The Once Upon a Time Map Book* (Candlewick, 2004)

Hutchins, Pat, *Rosie's Walk* (Simon & Schuster, 1968)